MONSANTO VS. THE WORLD

Monsanto, GMOs and Our Genetically Modified Future

Jason Louv

ULTRACULTURE PRESS

Published by Ultraculture Press
www.ultraculture.org
ultraculturegate@gmail.com

Cover photo via Flickr user Bdk.

Library of Congress Control Number: TBA

ISBN-13: 978-1484975909

10 9 8 7 6 5 4 3 2 1

Contents

Introduction

On March 28, 2013, President Barack Obama signed the short-term spending bill HR 933 into law to prevent the government from shutting down. Discreetly slipped inside the bill, however, was an additional rider—section 735, dubbed the Farmer Assurance Provision.

Environmental activists had their own nickname for it: the "Monsanto Protection Act." Outraged, they argued that the rider—which had been written by Senator Roy Blunt (R-Mo.), in collaboration with Monsanto itself[1]—would provide big agribusiness with immunity from judicial oversight. In effect, it would allow the US Department of Agriculture to approve of the planting of genetically modified crops *even if* the judiciary had declared them unsafe.

Monsanto, of course, tells a different story, stating that the rider is meant to allow farmers to plant crops without being disrupted for testing, which might financially cripple them.

"As we understand it," Monsanto spokeswoman Kelly J. Clauss responded to questions about the rider, "the point of the Farmer Assurance Provision is to strike a careful balance allowing farmers to continue to plant and cultivate their crops subject to appropriate environmental safeguards, while USDA conducts any necessary further environmental reviews."[2]

A letter provided directly from Monsanto's website, however, shows differently. Addressed to House Ranking Member Norm Dicks from Chairman Harold Rogers of the House

Committee of Appropriations, the letter states that the rider was constructed specifically to keep "activist groups" from disrupting the distribution of GMO crops by litigating against the USDA.[3]

While the Farmer Assurance Provision was passing through Congress, other storms were brewing: first it was announced that Kathleen Merrigan, one of the top officials at the USDA, was stepping down from her post as deputy secretary in an "abrupt" move.[4] Merrigan was the USDA's biggest champion of local and organic food, and had previously been outspoken about the big agribusiness lobby in Washington.

In the same weekend, Monsanto reached a settlement with its competitor DuPont, in which DuPont would pay Monsanto $1.75 billion in royalties for the right to use Monsanto's GMO technology.[5]

It was, from the perspective of Monsanto's critics, a *Godfather*-style coup—and for some Americans, the first time they had become aware of the existence of Monsanto, let alone the power the multinational had come to wield in the United States and worldwide since its move into agribusiness in the 1980s.

So who, exactly, is Monsanto? What are genetically modified crops, and are they dangerous? Who have we turned the future of our food—and by extension, our world—over to?

While a full exploration of the potential health concerns, complex scientific debate, and Kafka-esque political maneuverings over GMOs would fill many volumes and need constant updating, it is my hope that this short book can be of some service in understanding the details of this most crucial issue. I've also provided information on how to avoid GMO foods if you so choose.

Overall, I've sought to cut through the issues and show the bare details, whether that applies to Monsanto's policies, history and public statements, or to overstatements and unsupported claims by critics of Monsanto, so that the reader can be armed with more facts. I have included Monsanto's own responses to their critics throughout the text.

I hope this book is as eye-opening for you to read as it was for me to write.

Getting to Know Monsanto, Its Products and Its Policies

Monsanto (NYSE: MON) is a 111-year-old public multinational corporation—headquartered in Creve Coeur, Missouri—that deals in genetically modified organisms.

Previously a chemical and plastics company, Monsanto spearheaded the use of biotechnology in agriculture in the 1980s—it was the first private entity to genetically modify plants, in 1982, and is now one of the most prominent agricultural biotech companies in the world.

Among Monsanto's most successful products are *Roundup* herbicide and *Roundup Ready* seeds. Roundup is a weed-killer, and Roundup Ready crops are genetically engineered to be resistant to it—meaning that if you buy both, you can plant the Roundup Ready crops and then spray them with Roundup herbicide, and successfully kill only the weeds.

Monsanto also produces *Bt* seeds, which sprout crops that are genetically engineered to produce their own insecticide, in theory creating plants that don't need to be sprayed with additional chemicals (as they are already toxic to pests).

Founded in 1901 in St. Louis, Monsanto has a previous track record of highly toxic, often carcinogenic and in many cases now-illegal chemicals: they also brought the world the now-banned and carcinogenic DDT, the now-banned and carcinogenic PCBs, the controversial and potentially carcinogenic rBGH or bovine growth hormone, the controversial and allegedly carcinogenic artificial sweetener Aspartame, and the carcinogenic defoliant Agent Orange.

(Agent Orange was infamously used in the Vietnam War, resulting in an estimated 400,000 Vietnamese deaths, 150-500,000 Vietnamese children born with birth defects, and the spread of cancer or other debilitating conditions in nearly 40,000 American servicemen.)[6]

Monsanto was also tapped by the US Government to assist in the creation of the atomic bomb.[7]

Monsanto is currently one of the largest agribusiness corporations in the world, bringing in $11.82 billion a year in revenue (figure reported for the 2011 fiscal year). The company is at present headed by the Scottish-born business manager Hugh Grant, who has worked with Monsanto since 1981; his 2009 annual compensation for this position was $10.8 million.[8]

In addition, Monsanto is reportedly the most powerful agricultural lobbying interest in Washington,[9] spending $5.97 million on lobbying in 2012 and $6.37 million in 2011. According to the Center for Responsive Politics, Monsanto made over $1 million in campaign contributions between 2011-12: their three biggest recipients were Claire McCaskill (D-MO) at $32,524, Barack Obama (D) at $23,725 and Roy Blunt (R-MO) at $20,000.[10] It is worth repeating that Obama passed the Farmer Assurance Provision into law while Roy Blunt worked with Monsanto to craft the language of the rider;[11] McCaskill voted in favor of it.

What are Genetically Modified Organisms?

Genetically modified organisms are organisms that have genes mutated, deleted or even inserted into their own genetic makeup from a different species, via a process called *transgenesis.*

An example of a genetically modified crop is *Bt corn*, which has genetic material from a soil bacteria inserted into it so that the adult corn produces its own insecticide.

An example of a genetically modified animal is the "spider goat" created at Utah State University (not by Monsanto). These goats are engineered to produce spider silk proteins in their milk,[12] which can then be used to produce highly impermeable spider-silk body armor for the military.[13]

GMO proponents often claim that genetic modification has been ongoing since the beginning of human history. A crucial point to understand, however, is that what they're talking about is *plant breeding* and *hybridization,* not *transgenesis.*

Plant breeding *does* date back over 7,500 years, to when farmers in Mexico created maize by planting the crop *teosinte* and selecting for desirable genetic traits over the course of centuries.

Modern plant breeding (as in Norman Borlaug's famous dwarf wheat—see the section "GMOs and World Hunger," p. 31) involves cross-pollination of different plant species, or *hybridization.*

Genetic modification, however, is the direct genetic tampering with organisms and involves adding genetic sequences into plants that come either from other plants or even non-plant organisms.

The claim that genetic modification of crops is a millennia-old tradition is therefore only a half-truth, and one that is highly popular with pro-GMO lobbyists. Both Tony Blair and George W. Bush were fond of this rhetorical tactic while in office, as when Bush declared in 2001 that "for thousands of years, man has been utilizing and modifying biological processes to improve man's quality of life."[14]

Both Bush's father and Bush's cabinet, including Donald Rumsfeld and John Ashcroft, maintained high-level ties to Monsanto,[15] culminating in the post-Iraq war document Order 81, mandating that Iraqi farmers must purchase "registered" seeds, a massive windfall for the company.[16]

What is Transgenesis?

Transgenesis is the introduction of foreign genetic material into a living organism. Monsanto has most often achieved it by ballistic DNA injection or "gene gun," which blasts genetic material into undifferentiated plant cells in a petri dish—this destroys most of the cells, but usually leaves some that have successfully merged. Transgenesis can also be achieved using liposomes, plasmid vectors, viral vectors, pronuclear injection and protoplast fusion. Monsanto's current preferred method of transgenesis is using the bacteria *Agrobacterium tumefaciens* to infect plants with desired DNA traits.[17]

A controversial form of transgenesis is "pharming," slang for using the above techniques to insert genes that code for pharmaceuticals into living organisms.

Work on the technology for transgenesis began in the 1960s and has been in effective use since the early 1980s, when Monsanto achieved major breakthroughs in its implementation. Monsanto currently uses transgenesis technology to produce its highly successful *Bt* and *Roundup Ready* crops.

What is Bt Corn?

Bacillus thuringiensis is a soil bacteria that produces Bt proteins, which are toxic to insects (Bt bursts their stomachs). They have been used in insecticide sprays for decades—however, they're not effective against insects that hide inside the plants, and thereby manage to evade spraying.

Monsanto's answer was to insert Bt genes into both cotton and corn plants—in theory a major boost to crop yield, especially in the developing world. Bt allows both greater crop production (though this is a point of contention) and the decreased reliance on spray pesticides.

However, while insects have not yet shown resistance to Bt crops, that may only be temporary (as was the case with Monsanto's previous insecticide product DDT).

In 1999, a Cornell study (erased from the Web overnight during the course of the writing of this book; Wayback Machine link provided in footnotes)[18] found that Bt corn was toxic to Monarch butterfly caterpillars. This report initiated a storm of controversy over GMOs as well as how scientific data is reviewed, and the extent to which that process is mediated by non-objective concerns. Monsanto and other agricultural companies launched massive public relations campaigns to discredit the study, as well as founding a consortium (the Agricultural Biotechnology Stewardship Working Group) to fund counter-studies.[19] This helped establish a precedent that scientists reporting negative information about GMOs might face public attack. The USDA has since declared Bt corn safe for Monarch butterfly caterpillars.[20]

The environmental magazine *Orion* summarized the case thusly: "In the ongoing debate over the Cornell findings, the scientific process has been spun, massaged, and manipulated by the agricultural industry, the U.S. Department of Agriculture, the U.S. Environmental Protection Agency, and elements of the North American academic community. The process disregarded international scientific standards and has helped to make science the handmaiden of industrial agriculture. As a consequence of these irregular proceedings, the monarch–Bt corn debate risks losing sight of a larger, more serious issue: the real danger that genetically engineered crops will accelerate the industrialization of agriculture, human overpopulation, and the impoverishment of biological diversity."[21]

In 2000, a previously banned genetically modified corn called StarLink was found in taco shells; dozens had serious reactions to the GMO corn, including anaphylactic shock. Like Monsanto's Bt corn, StarLink had been engineered to produce insecticide, in this case by the biotech company Aventis. The EPA banned StarLink in 1998 for fear of its effect on humans, but it managed to find its way into the food supply anyway.[22]

In 2001, however, the Environmental Protection Agency (EPA) declared Bt cotton and corn safe.

What are Roundup and Roundup Ready?

Weed control is a major issue in farming of any size; weeds can decrease the size of crop yields and present a major headache. Farmers therefore must either manually weed (as in the developing world) or use herbicides to destroy weeds encroaching on their crops' space. However, making sure that herbicide only kills the weeds, and not crops, is a problem. Monsanto produces *Roundup*, or glyphosate, as a broad-use herbicide that works by breaking down an enzyme involved in amino acid synthesis, preventing affected plants from making proteins and thereby killing them.

In addition to Roundup, Monsanto produces *Roundup Ready* crops—soybean, maize, canola and cotton—which are genetically engineered to *resist* Roundup herbicide, so that these crops can be planted and then sprayed with Roundup, and survive while surrounding weeds do not.

The only competing technology is Bayer's *Liberty* herbicide (glufosinate), which also works by disrupting amino acid synthesis and leads plants to producing an overabundance of ammonia, killing them. Liberty is sold alongside *LibertyLink* crops, which are engineered to resist glufosinate. (LibertyLink was banned by the European Parliament on January 13, 2009.)

The EPA considers Roundup non-toxic to humans, even after a lifetime of consuming sprayed crops. However, they still mandate that products containing glyphosate be labeled with warnings against oral intake, for the use of protective clothing, and for users to not enter treated fields for at least four hours.[23]

Despite the EPA's stance, critics are skeptical of the use of Roundup Ready. Miguel Altieri, Professor of Agroecology at the University of California, Berkeley's Department of Environmental Science, Policy and Management wrote in 2007:

"Several scientists have argued that HRCs and Bt crops have been a poor choice of traits to feature this new technology given predicted environmental problems and the issue of resistance evolution. There is considerable evidence to suggest that both of these types of crops are not really needed to address the problems they were designed to solve. On the contrary, they tend to reduce the pest-management options available to farmers and there are many effective alternative approaches (i.e. rotations, polycultures, cover crops, biological control, etc.) that farmers can use to regulate the insect and weed populations that are being targeted by the biotechnology industry... I contend that GE crops will lead to further agricultural intensification and ecological theory predicts that as long as transgenic crops follow closely the pesticide paradigm, such biotechnological products will do nothing but reinforce the pesticide treadmill in agroecosystems."[24]

Many fear that both Bt and Roundup Ready crops additionally pose a risk to non-GMO crops, and will slowly edge out their non-GMO competitors by spreading into the wild or even by a process known as *gene transfer*, by which engineered genes might hop into non-GMO crops.

"GM crops do threaten organic agriculture in practical ways," states Mark Winston, Professor of Apiculture and Social Insects at Vancouver's Simon Fraser University. "Most notably, the mingling of GM genes with organic food seems inevitable, and while the extent of contamination may be low, tolerance by consumers may be zero. Also, the overuse and misuse of Bt varieties of GM crops by conventional farmers is likely, and the resulting induction of pest resistance would devastate organic growers."[25]

Norman C. Ellstrand, Professor of Genetics at the University of California, Riverside, confirms that gene transfer from transgenic crops may be an issue: "Gene flow from traditionally improved crops has on some occasions created problems by acting as a catalyst for weed evolution and by nudging wild relatives closer to extinction. We would expect the same for transgenic crops."[26]

In 2004, weeds that had developed resistance to Roundup appeared in genetically modified crop fields and began to spread throughout the American south; Roundup-resistant pigweed covered more than 100,000 acres in Georgia by 2009. The proposed solution was to use multiple herbicides, "thereby defeating the point of Roundup."[27]

In 2012, a French study[28] reported that mice fed Roundup Ready corn died two to three times more frequently than controls, developing tumors, liver and kidney problems, disabled pituitary glands and imbalanced sex hormones. This study was *widely* attacked on the Internet, including attacks from progressive sources.[29] It authors responded.[30]

The EPA has caught Monsanto-funded scientists falsifying data about glyphosate twice, in 1978[31] and 1991.[32]

What are Terminator Seeds?

GURT or genetic use restriction technology seeds, nicknamed "terminator seeds," are engineered to grow crops that produce sterile seeds, so that farmers would be prevented from saving seed from their own crops and thereby be forced to continue to buy seeds every time they replanted crops (saving or replanting Monsanto seeds already violates the company's user agreement with farmers, and has been repeatedly prosecuted). While the technology exists, and has been patented by Monsanto, it has never been publicly deployed, although rumors of its usage have caused protests in India and around the world.

Because of massive public outcry against their 1998 patent of terminator seeds, Monsanto backed down and publicly pledged not to actually deploy its patent in 1999. The company continues to display a promise on their website that "Monsanto has never developed or commercialized a sterile seed product. Sharing many of the concerns of small landholder farmers, Monsanto made a commitment in 1999 not to commercialize sterile seed technology in food crops. We stand firmly by this commitment. We have no plans or research that would violate this commitment in any way."[33] Later in the same pledge, however, they state that "If Monsanto should decide to move forward in the area of GURTs, we would do so in consultation with experts and stakeholders, including NGOs. Our commitment to protecting smallholder farmers and our promise not to commercialize sterile seed technology will carry forward with these developments, should they occur."[34]

Nature magazine reported in February 2013 that the concept of terminator technology may be regaining traction.[35]

What is Golden Rice?

Golden rice, of which Monsanto is a major patent-holder, is a GMO *Oryza sativa* rice variety that is golden yellow and engineered to produce large quantities of Vitamin A, deficiency in which is a major source of infant mortality in the developing world.[36] It has not yet been deployed; however, the Bill and Melinda Gates Foundation, which owns over 500,000 shares in Monsanto (worth $23 billion), has provided over $20 million for its continued development.[37] Critics, such as Dr. Vandana Shiva, have suggested that the benefits of Golden Rice are minor compared to the threat posed to farmers who will have to submit to the policies of the seeds' patent holders, instead arguing simply for eating non-monoculture vegetables and greens to get sufficient Vitamin A.[38]

Monsanto's Aggressive Litigiousness

Monsanto has been criticized for its litigious attitude towards small farmers. Since the 1990s, the company has filed 145 lawsuits against individual farmers, either for patent infringement or breach of contract. A common trope among activists is that Monsanto's seeds will blow into farmers' land from neighboring GMO crops and take root, and that this will be enough legal ground for Monsanto to sue the farmer. This stems from the case of Percy Schmeiser, a farmer in Saskatchewan, Canada who saved Monsanto seeds that had blown into his farm in 1997 and replanted them in 1998. His 1998 planting, however, was found to be 95-98% pure Roundup Ready canola, for which the Canadian Supreme Court could find no explanation other than intentional piracy—however, the court also found that Schmeiser had not in any way profited from re-sowing Monsanto's seeds. Schmeiser won a partial victory in the Supreme Court, and was not forced to pay damages.[39]

When farmers purchase seed from Monsanto, they sign an agreement that they will not save or replant Monsanto seeds. Monsanto actively polices their copyrights, and encourages farmers who witness other farmers saving or replanting their seeds to inform on their neighbors directly to a Monsanto 800 number.[40] They also, according to Monsanto itself, employ private investigators to locate farmers who may be breaking their agreements with the company.[41]

In 2013, the Supreme Court reviewed a case expected to set a precedent on whether or not farmers can legally save seeds (*Vernon Hugh Bowman v. Monsanto Company*).[42] Associate Justice Elena Kagan, who previously ruled in favor of Monsanto's GMO alfalfa (see "What is the Revolving Door?," p. 22) found in favor of Monsanto on May 13, 2013.

Concern about Monsanto's active policing of its interests extends beyond farmers: a 2010 article in *The Nation* alleged that Monsanto had hired the services of the infamous private security company Blackwater (now renamed Xe Services) to "infiltrate activist groups organizing against the multinational biotech firm."[43] Monsanto denies this allegation on their website, stating that they only retained the related group TIS (Total Intelligence Solutions) to provide "Monsanto's security group with reports about activities or groups that could pose a risk to the company, its personnel or its global operations," and did not infiltrate activist groups. TIS is owned by The Prince Group, which also owns Blackwater (Xe); Monsanto states that "prior to retaining TIS, Monsanto specifically enquired about and was informed that TIS was a completely separate entity from Blackwater."[44]

According to 2007 documents released by WikiLeaks as part of Cablegate, the United States embassy in Paris, France advised Washington to initiate "military-style trade war" (*The Guardian*'s phrase)[45] against any country in the European Union that opposed the introduction of GM crops.

The request came from US ambassador Craig Stapleton—Stapleton co-owned the Texas Rangers baseball team with George W. Bush from 1989-1998;[46] his wife, Dorothy Walker Stapleton, is George W. Bush's cousin.[47] Stapleton's leaked cable, which was transmitted to Washington after France began moves to ban Monsanto GM corn, reads:

"Country team Paris recommends that we calibrate a target retaliation list that causes some pain across the EU since this is a collective responsibility, but that also focuses in part on the worst culprits.

"The list should be measured rather than vicious and must be sustainable over the long term, since we should not expect an early victory. Moving to retaliation will make clear that the current path has real costs to EU interests and could help strengthen European pro-biotech voices."

Other WikiLeaks-released cables, as reported by *The Guardian*, "show US diplomats working directly for GM companies such as Monsanto"; that US diplomats had been active in pushing support for GM crops at the Vatican due to the resistance of Catholic bishops to GM crops in the developing world; and even that Spain and the United States had worked together to "persuade the EU not to strengthen biotechnology laws."[48]

Indian Farmer Suicides

Monsanto has been blamed for the rash of farmer suicides in India—even Prince Charles stated publicly and stridently in 2008 that "I blame GM crops for farmers' suicides."[49] 17,500+ farmers committed suicide per year in India from 2002-06;[50] according to the Indian government, 200,000 farmers committed suicide between 1997-2010.[51] The suicides were first reported in the western state of Maharashtra (India's wealthiest state), though were initially denied by the government. Though figures have been debated, a 2011 study by the *Lancet* found that the numbers might actually have been under-reported (in many cases because suicide figures relied on local ledgers for criminal activity—as suicide is heavily socially stigmatized in India, many may have gone unreported),[52] suggesting that there were 19,000 farmer suicides in 2010.[53]

According to the BBC, "Campaign groups claim the suicides have been caused by food speculators manipulating cereal prices, and GM companies who are selling expensive cotton seeds and fertilizers. They say that in order to buy GM seeds, some farmers get into unmanageable debt. Others are crippled by fluctuations in food prices. And when the going gets too tough some decide the only way out is to take their own lives."[54]

Critics, however, have suggested that farmer suicides in India are not statistically higher than any other group across India as a whole.

"The farmer suicides started in 1997," seed activist Dr. Vandana Shiva told CNN in 2010. "That's when the corporate seed control started, and it's directly related to indebtedness, and indebtedness created by two factors linked to globalization. The combination is unpayable debt, and it's the day the farmer is going to lose his land for chemicals and seeds, that is the day the farmer drinks pesticide. And it's totally related to a negative economy, of an agriculture that costs more in production than the farmer can ever earn."[55] Shiva singles out Monsanto's Bt cotton in particular for the rash of suicides.[56]

Monsanto has stated on their corporate blog "Beyond the Rows" that "Higher suicide rates among Indian farmers long predate the introduction of biotech cotton in India." Monsanto points to Bt cotton as a source of economic boom for India, which is the world's fastest growing consumer of biotech, instead blaming farmer debt on a variety of socioeconomic factors—including bad business skills, buying imitation Monsanto seeds and debt from "endowment obligations for the marriage of daughters."[57]

What is the Revolving Door?

"Revolving door" is a general political term, not limited to agribusiness, describing the rotation of personnel between legislative roles in the government and into lobbying roles in the industries affected by that legislation, often back and forth over a period of many years.

Of particular importance to the public dialogue on Monsanto, environmentalists and progressives have raised red flags over Obama's appointment of Michael R. Taylor to the position of Deputy Commissioner of Foods at the FDA.

Taylor joined the FDA as a staff attorney in 1980, after which he went into private practice at the law firm King & Spalding, of which Monsanto was a client. Returning to the FDA as Deputy Commissioner for Policy in 1992, he allegedly co-wrote and signed the Federal Register notice that milk from cows treated with bovine growth hormone (produced by Monsanto) did not have to be labeled, as well as allegedly ensuring that the FDA would not interfere with the production of GMO foods. During this time he was charged by activist Jeremy Rifkin as having a conflict of interest over the approval of rBST (Bovine somatotropin); the FDA replied that Taylor was in violation of no laws and "adhered to all applicable ethical standards."[58]

After serving at the USDA, Taylor was appointed Monsanto's vice president for public policy in 1998. On January 13, 2010, Obama appointed him as Deputy Commissioner of Foods at the FDA. Critics point to this as a particularly egregious example of the agribusiness revolving door, by which personnel move between holding positions at Monsanto, the FDA, the EPA and the USDA on an interchangeable basis.

Supreme Court Justice Clarence Thomas was employed by Monsanto in the 1970s,[59] and Associate Justice Elena Kagan has previously gone to bat in favor of Monsanto's GE alfalfa[60] (though has not held a position with Monsanto).

Obama also appointed Monsanto lobbyist Islam Siddiqui to the post of Chief Agricultural Negotiator in the Office of the United States Trade Representative in 2011. In 1998, as Under Secretary for Marketing and Regulatory Programs at the USDA, Siddiqui wrote the standards for organic food labeling that permit GMO *and* irradiated food to be labeled organic.[61] Obama-appointed Secretary of Agriculture Tom Vilsack is a major proponent of GMO crops and deeply linked to the biotech lobby; as governor of Iowa in 2005, he created a seed pre-emption bill that blocked local communities from regulating GMOs.[62] (Obama's cabinet is by no means unique; Monsanto also maintained intimate ties with the Bush and Clinton cabinets.)

Kathleen A. Merrigan, who left office at the USDA the same weekend the "Monsanto Protection Act" was passing through Congress, had this to say in 2007 about the biotech lobby:

"In 1987, I was hired as a congressional staff aide for the U.S. Senate Agriculture and Judiciary Committees to manage emerging issues surrounding agricultural biotechnology... During my first week on the job, I was flown to the headquarters of a major biotechnology company and briefed on the remarkable science that the company was developing.

It was explained that the technology would help to feed the world while simultaneously reducing environmental degradation associated with conventional production methods. The industrial lobbyists worked diligently to ensure that I understood that biotech is the great elixir that would not only end starvation, but also cure cancer and infuse great wealth into the American economy. However, they did raise a caution and urged my cooperation. Biotech would only succeed if the government did not overreact to unscientific alarmists who were using rhetorical powers and subversive tactics against the industry. From the onset on my Senate career, industry and government leaders alike advised me that if biotech were to fulfill its promise, the appropriate roles for the government were to (1) recognize that biotech is not remarkably different from conventional breeding and to regulate it accordingly; (2) invest in biotech research and education and remove disincentives to commercialization; and (3) provide adequate property rights to reward and encourage invention. Then and now, these three principles undergird American biotech policy... rather than try to change these principles... advocates for system change should work within them."[63]

Change from within was, it would seem, Merrigan's approach until her "'abrupt' departure"[64] from the USDA as the Farmer Assurance Provision was being approved.

Henry Miller, head of biotechnology at the FDA from 1979 to 1994, stated in 2001 that government agencies did "exactly what big agribusiness had asked them to do and told them to do."[65]

Monsanto's Greatest Hits

Scientific debate over Monsanto's Bt, Roundup and other GMO products is furious, well-funded on one side, and labyrinthine. It is worth noting that the current mainstream scientific consensus—whatever one thinks of the motivation and funding of that consensus—is that Monsanto's products are safe and beneficial solutions for world hunger.

However, this is not the case for many of their previous products. Let's take a look at Monsanto's prior track record of chemical products in the 20th century, the toxicity of many of which has been settled in court.

DDT

DDT was a Monsanto-produced insecticide in use until environmentalist Rachel Carson's 1962 book *Silent Spring* caused a massive public backlash that led to its ban in the United States in 1972. Up until 1962, pesticides had not been regulated in the United States; the controversy over DDT was the impetus for the creation of the Environmental Protection Agency by Richard Nixon in 1970. (William Doyle Ruckelshaus, the man Nixon appointed as the first director of the EPA, went on to serve on Monsanto's board of directors, by the way.)

DDT was initially used to combat typhus and malaria, with major success, but soon led to insect resistance (a major concern environmentalists also have in regards to Roundup)—in many cases leading to an even greater problem.[66] DDT was additionally linked to cancer (DDT is directly genotoxic to human beings, disrupting enzyme production and the endocrine and reproductive systems) and the destruction of wildlife, also potential concerns with Roundup and Roundup Ready crops.

Many countries still use DDT for malaria control.

PCBs

Polychlorinated biphenyls or PCBs are organochlorides that were used as insulating fluid for industrial machinery, as well as in PVC insulation for wires.

They were found to be persistent organic pollutants and environmental toxins, carcinogenic to animals and likely humans—especially at cause for non-Hodgkin Lymphoma.

PCBs were banned by the United States Congress in 1979; prior to this, Monsanto produced *99 percent* of the PCBs in use in the United States.

The dangers of PCBs, however, were known as far back as *1897,* when they were observed producing grotesque blackheads, hard cysts and pustules that covered the bodies of workers exposed to them. After another outbreak in 1934, Monsanto shaved nineteen rabbits and covered them with PCBs, producing ulcerous lesions across their bodies. The results were unmistakable.

Shortly after this, Monsanto was sued by workers exposed to PCBs in its Anniston, Alabama facility. The company's response was to build ventilation, provide fresh clothing and cold cream each day, and construct areas for workers to bathe and swab themselves with alcohol after each shift—once they were off the clock.

Monsanto *never* informed the public of the dangers of PCBs, and would not cease their manufacture until ordered to do so by the federal government over four decades later, spending that time actively convincing their customers of the safety of PCBs.[67]

Monsanto was sued in 2002 by over 20,000 residents of Anniston for knowingly dumping mercury and PCBs into the local creeks for decades (45 tons of PCBs in 1969 alone), and burying millions of pounds of PCBs in open landfills in the surrounding area. The company was ordered to pay over $700 million in settlements.[68]

Bovine Growth Hormone

Approved by the FDA in 1993, rBGH or recombinant bovine growth hormone—marketed as Posilac in the United States—is a drug designed to turbocharge cow milk production.

Milk from cows treated with rBGH contains 70% higher levels of IGF-1 or insulin growth factor; the British journal *The Lancet* reported in 1998 that higher blood levels of IGF-1 in premenopausal women creates a *seven times higher* chance of developing breast cancer.[69] rBGH milk has since been banned in the European Union, Canada, Japan, Australia and New Zealand.

Yet both Monsanto and the FDA maintain the party line that there is no difference between normal milk and milk from cows treated by rBGH.

Meanwhile, packaging on Posilac mentions that it may cause "decreases in gestation length and birth weight of calves," "increased risk of clinical mastitis (visibly abnormal milk)... The number of cows effected with clinical mastitis and the number of cases per cow may increase. In addition, the risk of subclinical mastitis (milk not visibly abnormal) is increased," "higher milk somatic cell counts,"[70] and so forth.

Monsanto estimated in 1998 that 30% of US cows were on Posilac.[71]

In 2008, Monsanto sold the entirety of their rBGH business to pharmaceutical multinational Eli Lilly for $300 million plus additional consideration.

Aspartame

Aspartame is a Monsanto-developed artificial sweetener found in a wide variety of foods, notably diet soda, approved for use by the USDA in 1981.

Aspartame is widely held in public opinion to be both carcinogenic and neurotoxic, in some cases due to chain e-mails.[72] However, a 2007 medical review found it safe for public consumption, and it remains in wide use despite public skepticism.[73]

Agent Orange

During the Vietnam War, Monsanto and Dow Chemical produced the defoliant Agent Orange for the US Department of Defense, a mixture of equal parts 2,4,5-T and 2,4-D. Between 1962 and 1971, 20 million gallons of Agent Orange and related compounds were sprayed over Vietnam, Laos and Cambodia to kill trees and vegetation so that Viet Cong soldiers could be shot from the air, and so that peasants, now unable to support themselves on farming, would flee the countryside and be forced into US-controlled urban areas.[74]

Agent Orange affected between 3-4.8 million Vietnamese people, resulting in a reported 400,000 deaths and between 150-500,000 children born with birth defects. The defoliant destroyed 17.8% of the Vietnamese environment and, according to the Red Cross of Vietnam, created health problems or disabled 1 million people.[75]

Agent Orange also affected the American servicemen who deployed or were exposed to it, causing, according to Wikipedia, "throat cancer, acute/chronic leukemia, Hodgkin's lymphoma and non-Hodgkin's lymphoma, prostate cancer, lung cancer, colon cancer, soft tissue sarcoma and liver cancer."[76] The Department of Veterans Affairs had received claims from 39,419 soldiers adversely affected by Agent Orange by 1993.[77]

Multiple lawsuits have been filed against Monsanto, Dow and Diamond Shamrock over Agent Orange since 1978. A 1984 class-action lawsuit saw a $180 million combined payout from seven chemical companies, of which 45% was paid by Monsanto—the out-of-court settlement outraged Vietnam veterans, who wanted to see the action settled in court and the chemical companies held responsible for their product. Once settled, the final payout to affected veterans came down to only $12,000 spread out over ten years, *merely $1,200 per year*. Furthermore, accepting the payout would make veterans ineligible for state assistance, pensions or food stamps, which in many cases were worth more than $1,200 a year, meaning that many were forced to not accept anything at all.[78]

In 2004, Monsanto spokeswoman Jill Montgomery stated that "We are sympathetic with people who believe they have been injured and understand their concern to find the cause, but reliable scientific evidence indicates that Agent Orange is not the cause of serious long-term health effects."[79]

The Atomic Bomb

Monsanto executive Charles Allen Thomas was tapped to extract and purify the radioactive element polonium, which was used in the Manhattan Project to construct the neutron generating devices that triggered the detonation of the atomic bombs used in Hiroshima and Nagasaki.[80]

Whatever one's political view of these events, I believe the lethal, destructive and carcinogenic properties of atomic bombs need no further elaboration.

GMOs and World Hunger

Monsanto's primary argument for Genetically Modified Organisms is that they may provide a solution to world hunger.

GMO proponents often point to the success of Nobel laureate Norman Borlaug (1914-2009), sometimes referred to as "The Man Who Saved a Billion Lives," who led the Green Revolution by introducing high-yield varieties of wheat into the developing world starting in the 1960s, increasing food production in India, Pakistan, Asia and Africa.

Of critics of his methods, who state that he brought monoculture to areas that previously relied on subsistence farming, and forced the agenda of agribusiness and first world capitalism into developing world nations, Borlaug stated that "some of the environmental lobbyists of the Western nations are the salt of the earth, but many of them are elitists. They've never experienced the physical sensation of hunger. They do their lobbying from comfortable office suites in Washington or Brussels. If they lived just one month amid the misery of the developing world, as I have for fifty years, they'd be crying out for tractors and fertilizer and irrigation canals and be outraged that fashionable elitists back home were trying to deny them these things."[81]

It is important to note, however, that Borlaug's wheat was developed by *cytogenic hybridization techniques*, meaning crossbreeding different strains of wheat and selecting for favorable traits. They were not developed by *transgenesis*—Monsanto did not begin work on transgenic crops until the 1980s, when they achieved their initial successes with the new method that would lead to Bt corn, Bt cotton and Roundup Ready crops.

(Critics, frostily, have stated that Borlaug's dwarf wheat is a major contributor to world overpopulation.[82] The number of people in the world has increased by *over four billion* since the Green Revolution sparked by Borlaug's work; the adoption of transgenic wheat *might* create an even greater population explosion. We are well into *highly* uncomfortable ethical territory at this point.)

Wheat was the *last* major crop to be modified by transgenic methods, in 1992,[83] and transgenic wheat has not yet been widely adopted. Borlaug's success, good or bad, has *nothing* to do with Monsanto's current transgenic Bt and Roundup Ready products.

As for the latter technology, *Scientific American* has reported that claims that transgenic GMO crops produce larger yields are false. As they stated on April 16, 2009:

"Proponents argue that GM crops can help feed the world. And given ever increasing demands for food, animal feed, fiber and now even biofuels, the world needs all the help it can get.

"Unfortunately, it looks like GM corn and soybeans won't help, after all.

"A study from the Union of Concerned Scientists[84] shows that genetically engineered crops do not produce larger harvests.

Crop yield increases in recent years have almost entirely been due to improved farming or traditional plant breeding, despite more than 3,000 field trials of GM crops.

"Of course, farmers have typically planted, say, GM corn, because it can tolerate high doses of weed-killer. And the Biotechnology Industry Organization argues that GM crops can boost yields in developing countries where there are limited resources for pesticides.

"But it appears that, to date, traditional plant breeding boosts crop yields better than genetic modification. Those old farmers were on to something."[85]

As of 2010, there were an estimated 925 million hungry people in the world, mostly in Asia and Sub-Saharan Africa.[86] In November 2012, the journal *Science of the Total Environment* released a study that suggested a "novel" approach to remedying this figure: cutting waste in the current food production chain by only 50%, they found, would feed over 1 billion people.[87]

GMOs and Public Health

The party line held by Monsanto, the FDA and the USDA is that there is *no difference* between natural and genetically modified foods. (Author Kathleen Hart, for instance, was told by an FDA spokesman that there is "no substantial difference between genetically engineered foods and their conventional counterparts," and was given similar lines by Monsanto and the USDA, who dismissed the European policy of labeling GMO food as "based on emotion.")[88]

This party line was allegedly constructed by Michael R. Taylor in 1991 while serving as Deputy Commissioner for Policy at the FDA. He had previously worked in private practice at a law firm that retained Monsanto as a client; post-1991 he went on to work for the USDA and then for Monsanto directly as Vice President for Public Policy, before being re-appointed to the FDA by Obama in 2010 as Deputy Commissioner for Foods.

However, many food products already contain voluntary notifications that they refuse to use GMOs or rBGH. This speaks to a growing public awareness and concern about the safety of these foods.

Consumer advocates and activists have blamed GMO crops for inflammation, allergic reactions (including priming the body to have allergic reactions to non-GMO foods), immune responses, damaged intestines, cancer, liver and kidney problems, reproductive problems, infant mortality, birth defects, sterility and death. They have also stated that Bt corn leaves Bt proteins within the digestive system that continue to manufacture pesticide within the consumer's own intestinal flora long after the actual GM food has been digested.[89]

Many of these claims are exceedingly hard to validate. Most of the studies that have raised these claims have been shredded by the scientific community (for instance, the French study discussed in the section on Roundup); the originating scientists and food activists usually claim that this is due to lack of government oversight combined with wide-scale funding by agribusiness for studies to prove the safety of GMOs and to discredit dissenting claims. Yet this does not change the fact that, at present, the scientific consensus remains that GMO foods pose no dangers.

However, it is worth nothing that the USDA currently *does not* require biotech companies to do premarket safety or allergen testing, despite public urging by the American Medical Association to do so.[90]

Another major concern with GMOs is *gene transfer,* the fear that genes from GMO crops will migrate into non-GMOs or animals—even humans. A particularly grim scenario would be an antibiotic-resistant gene migrating into a pathogen within the stomach of somebody who has consumed GMO food, potentially producing a pandemic. 2004 research by the journal *Nature Biotechnology* showed that this was not a danger; however, this remains the only major study yet conducted.[91]

A 2011 Canadian study reported that Cry1Ab, an insecticidal protein found in Bt and Roundup crops, was found in the cord blood of fetusus and the blood of the mothers who had consumed the crops.[92] This article also came under massive attack for its methodology—in addition, no dietary evidence was provided for any of the subjects.[93]

This means that the scientific argument for the toxicity of GMOs is on shaky ground. However, as critics point out, this may be more indicative of a stacked deck than actual lack of risk.

As Dr. Michelle Marvier, Chair of the Department of Environmental Studies and Sciences at Santa Clara University pointed out in 2007 (specifically of the potential environmental, not health, risks of GMOs):

"For the majority of US petitions for deregulation, the approaches used to support the environmental safety of GE crops rely primarily on the following:

"1. Conducting small (meaning poorly replicated) trials to test for effects and

"2. Citing published and unpublished studies or using letters from expert scientists to establish an absence of risk...

"A recent, hard-hitting report by the National Research Council[94] recommended that the USDA should improve the rigor and transparency of its decision-making process... serious consideration of the environmental risks associated with GEOs is no more 'antitechnology' than requiring safety screening for new drugs is 'anti-medicine.' The fact that a particular drug may save millions of lives does not eliminate the need for extensive investigation of the drug's side-effects... After all, if a newly released drug turns out to be dangerous, we can simply take the new drug off the market. In contrast, if genes from problematic GEOs find their way into nature, they cannot be so easily recalled."[95]

Research Professor Dr. Lincoln P. Brower, a distinguished elder of the genetics debate, has painted an even more stark picture of a GMO future: "Looking beneath the purported advantages of the new GMO technology to agriculture and corporate profits," he writes, "an alternative view is that these corporations are converting the natural world into a biologically impoverished planet massively overpopulated by a single species: Homo sapiens. The sweeping extent of this technology can be seen in chicken factories that sit in the middle of vast cornfields, devoid of all native plants. The rich web of life that formerly occupied this prairie community has been reduced to an industrial food chain that has only three links: sunlight to corn, corn to chickens and chickens to humans."[96]

In 1998, citizens in Scotland, England and Ireland tore up GMO crops planted by Monsanto, Aventis and Novartis. Even Prince Charles joined their cause, stating that agribusiness was conducting a "gigantic experiment... with nature and the whole of humanity which has gone seriously wrong. What we should be talking about is food security not food production—that is what matters and that is what people will not understand."[97] In response, Monsanto launched a $1.5 million PR campaign to convince the British public to relax. Yet protests soon spread across the European continent, and on May 26, 1998, the European Union penned laws requiring genetically engineered foods to be labeled. In 2011, Hungary destroyed every crop field it found to contain GMOs—over 1,000 acres; distribution of GMOs is illegal in that country.[98]

Yet no such drastic measures have occurred in America. In the United States, debate over GMOs is often the province of wealthy leftists, for whom purchasing highly expensive organic and non-GMO food at Whole Foods, farmer's markets or other sources is an option, even a marker of social status or upper middle class aspiration. Such markers of privilege are often simply outside of the access of those whose budget or locale makes shopping at Wal-Mart and dining out at McDonald's more realistic—even a privilege in and of itself; let alone those unable to afford food at all. (Self-cultivation of crops *may* somewhat dodge this class divide.) That the organic and GMO debate is deeply intertwined with issues of class and privilege is an uncomfortable truth, and not one that is widely addressed.

In many ways, the cultural debate over GMOs suffers because it is fought between extremes—spokespeople for agribusiness and environmental activists—with little to no middle ground. It is telling, for instance, that when the controversy over the "Monsanto Protection Act" erupted, news sources and media consumers initially had to look either to press release attacks from environmental groups or public statements from Monsanto itself. When I was assembling my own blog posts on the subject (as HR 933 was passing through Congress), the *only* mainstream source I found on the Farmer Assurance Provision was a minor report from NPR, a damning statement about the state of journalism; that has since been rectified, though of course alongside more level-headed journalism the Internet has also cluttered up with more extremist and half-formed opinions.

Robert Paarlberg, Adjunct Professor of Public Policy at Harvard's John F. Kennedy School of Government, who specializes in global food and agricultural policy, had this to say about the extremist polarization in 2002:

"Tragically, the leading players in this global GM food fight—US based industry advocates on the one hand and European consumers and environmentalists on the other—simply do not reliably represent the interests of farmers or consumers in poor countries... The subsequent public debate naturally deteriorates into a grudge match between aggressive corporations and their most confrontational NGO adversaries. Breaking that paralysis will require courageous leadership, especially from policymakers in developing countries. These leaders need to carve out a greater measure of independence from the GM food debate in Europe and the United States. New investments in locally generated technology represent not just a path to sustainable food security for the rural poor in these countries. In today's knowledge-driven world such investments are increasingly the key to independence itself."[99]

How to Avoid GMO Food

It is unlikely that GMO food will be going away any time soon. It is also unlikely, for reasons discussed above, that scientific consensus will declare GMOs harmful to consumers.

This leaves the awareness of the issues surrounding GMO food, and the decision whether or not to purchase and consume it, down to the individual.

Luckily, there a multitude of steps you can take to eliminate GMO foods from your own diet and that of your family if you so choose.

1. Know what foods are most commonly genetically modified. These foods in particular, unless labeled 100% organic, are often GMOs:

• Corn and corn-derivative products like high-fructose corn syrup. Will carry either Bt proteins or be Roundup Ready engineered.

• Soybeans and soy products (tofu, soy milk). Roundup Ready.

• Canola (usually as canola oil). Roundup Ready.

• Sugar Beets. Roundup Ready.

• Dairy Products. Produced by cows given rBGH and fed on GMO corn, grains or hay.

• Sugar. Often comes from Roundup Ready sugar beets.

- Pre-prepared foods containing any of the above ingredients or derivatives.

- Hawaiian Papaya

- Zucchini

- Yellow Crookneck Squash

2. Avoid cooking with canola, corn, soy or cottonseed oil. These are all likely made with GMO crops. Butter may also be from cows raised on GMO feed or given rBGH.

3. If you live in Europe, buy non-GMO. If you live in the United States, buy 100% Organic.

Foods labeled "organic" in the United States or Canada are not necessarily GMO-free. However, foods labeled specifically "100% Organic" are by law. You are liable to spend considerably more for these foods, an unfortunate economic reality at present, but they should be GMO-free.

4. If you eat beef, buy 100% grass-fed. Finding a 100% label is once again critical here, in order to make sure that the cows weren't exposed to GMO feed at any point in their lives. 100% grass fed beef is expensive, but can often be bought cheaply in bulk directly from family farms over the Internet and stored in a freezer. It also tastes *significantly* better than factory farmed beef—you won't want to go back.

5. Learn fruit and vegetable encoding. Fruit and vegetable stickers carry four-to-five digit numbers. A four-digit number

means the item is non-GMO. A five-digit number beginning with an 8 is GMO (though *identification of GMO foods is optional*). A five-digit number beginning with a 9 means the item is organic.

6. Avoid processed or pre-prepared foods. As mentioned above, it can be immensely difficult to know what you're getting, and processed food is bad for you anyway. Buy whole foods instead. Farmer's markets may be good places to shop, as you can often get non-GMO food there and have conversations with the farmers selling the food on the spot, getting an education on food in the process. Shopping at local co-ops may be another good option.

7. Team up. Because shopping for non-GMO food is a hard task, requiring extra research, awareness, preparation, travel and money, it may be highly beneficial to develop a group of neighbors or to a join a local co-op to turn it into a team effort, sharing duties and information. Such a group could also easily develop into a communal gardening or crop-planting effort, providing not only healthy food, but a sense of community and mutual aid.

8. Grow your own food—and plant non-GMO seeds. This can be challenging, as seeds are not labeled as GMO or non-GMO in the United States. However, *open pollinating* and *heirloom* seeds are more likely to be non-GMO. Search online for non-GMO seeds—a wide range of options for buying can readily be found.

I wish you the best of luck in whatever choice you make for yourself and your family.

Endnotes

[1] Knowles, David. "Opponents of genetically modified organisms in food, or GMOs, rail against provision that would limit the courts' ability to stop food producer Monsanto from growing crops later deemed potentially hazardous." *New York Daily News,* March 25, 2013. http://www.nydailynews.com/news/national/food-oversight-curbs-spending-bill-outrage-article-1.1298967

[2] "Monsanto Statement Regarding Farmer Assurance Provision in H.R.933." Monsanto.com. http://www.monsanto.com/newsviews/Pages/statement-regarding-farmers-assurance-provisions.aspx

[3] Rogers, Harold, Hon. Letter to Honorable Norm Dicks. June 12, 2012. Monsanto.com. http://www.monsanto.com/newsviews/Documents/Letter%20to%20House%20for%20Section%20733.pdf

[4] Laskawy, Tom. "Sustainable food loses its biggest champion in Washington, D.C." *Grist*, March 25, 2013. http://grist.org/food/sustainable-food-loses-its-biggest-champion-in-washington-d-c/

[5] Gillam, Carey. "Monsanto, DuPont strike $1.75 licensing deal, end lawsuits." Reuters. March 26, 2013. http://www.reuters.com/article/2013/03/26/us-monsanto-dupont-gmo-idUSBRE92P0IK20130326

[6] http://en.wikipedia.org/wiki/Agent_Orange

[7] http://en.wikipedia.org/wiki/Dayton_Project

[8] http://en.wikipedia.org/wiki/Monsanto

[9] http://www.opensecrets.org/lobby/indusclient.php?id=A07&year=2012

[10] http://www.opensecrets.org/orgs/summary.php?id=D000000055

[11] (Knowles.)

[12] "The goats with spider genes and silk in their milk," BBC News. http://www.bbc.co.uk/news/science-environment-16554357

[13] Chan, Amanda. "'Bullet-Proof Skin', Made With Spider Silk and Goat's Milk, Created by Scientists." *Huffington Post,* August 18, 2011. http://www.huffingtonpost.com/2011/08/18/bullet-proof-skin-spider-silk_n_930389.html

[14] Federal Register Volume 66, Number 98. May 21, 2001. pp. 28045-28046 http://www.gpo.gov/fdsys/pkg/FR-2001-05-21/html/01-12966.htm

[15] "Continuing the Green Revolution: The corporate assault on the security of the global food supply." Massachusetts Institute of Technology: *The Thistle.* Vol. 13, No. 4. June/July 2001. http://web.mit.edu/thistle/www/v13/4/food.html

[16] Roberts, Jeanne. "The Real Victor in Iraq: Monsanto," in *The Panelist.* http://www.thepanelist.net/opinions-culture-10084/1252-the-real-victor-in-iraq-monsanto

[17] Palmer, Roxanne. "GMO Health Risks: What the Scientific Evidence Says." *International Business Times,* March 30, 2013. http://www.ibtimes.com/gmo-health-risks-what-scientific-evidence-says-1161099

[18] "Toxic pollen from widely planted, genetically modified corn can kill monarch butterflies, Cornell study shows." Ithaca: Cornell University. May 19, 1999. http://web.archive.org/web/20130323114050/http://www.news.cornell.edu/releases/May99/Butterflies.bpf.html

[19] Brower, Lincoln. "The Monarch and the Bt Corn Controversy." *Orion*, Spring 2001. http://www.orionmagazine.org/index.php/articles/article/85

[20] The Pew Trust. *Three Years Later: Genetically Engineered Corn and the Monarch Butterfly Controversy.* http://www.pewtrusts.org/uploadedFiles/wwwpewtrustsorg/Reports/Food_and_Biotechnology/vf_biotech_monarch.pdf

[21] (Brower.)

[22] Hart, Kathleen. *Eating in the Dark: America's Experiment With Genetically Engineered Food.* New York: Pantheon Books, 2002.

[23] Roundup Pro® Herbicide Complete Directions for Use. EPA Reg. No. 524-475. http://www.afpmb.org/sites/default/files/pubs/standardlists/labels/6840-01-108-9578_label_roundup_pro.pdf

[24] Altieri, Miguel A. "Transgenic Crops, Agrobiodiversity, and Agroecosystem Function," in Tayor. Iain E. P., ed., *Genetically Engineered Crops: Interim Policies, Uncertain Legislation.* Binghamton: The Haworth Press, 2007. p. 39

[25] Winston, Mark L. *Travels in the Genetically Modified Zone.* Cambridge: Harvard University Press, 2002. p. 171

[26] Ellstrand, Norman C. *Dangerous Liasons? When Cultivated Plants*

Mate With Their Wild Relatives. Baltimore: Johns Hopkins University Press, 2003. p. 186

[27] Nestle, Marion. *Safe Food: The Politics of Food Safety*. Berkeley: University of California Press, 2010. p. 279

[28] Séralini GE, Clair E, Mesnage R, Gress S, Defarge N, Malatesta M, Hennequin D, de Vendômois JS. "Long term toxicity of a Roundup herbicide and a Roundup-tolerant genetically modified maize." http://www.ncbi.nlm.nih.gov/pubmed/22999595

[29] Pinder, Terry. "Monsanto is a bad corporation but that GM Maize study is dodgy as heck." *Daily Kos*, September 20, 2012. http://www.dailykos.com/story/2012/09/20/1134246/-Monsanto-is-a-bad-corporation-but-that-GM-Maize-study-is-dodgy-as-heck

[30] Séralini GE, Clair E, Mesnage R, Gress S, Defarge N, Malatesta M, Hennequin D, de Vendômois JS. "Answers to critics: Why there is a long term toxicity due to a Roundup-tolerant genetically modified maize and to a Roundup herbicide." http://www.ncbi.nlm.nih.gov/pubmed/23146697

[31] Schneider, K. "Faking It: The Case Against Industrial Bio-Test Laboratories." *The Amicus Journal*, Spring 1983. http://planetwaves.net/contents/faking_it.html

[32] EPA FY1994 Enforcement and Compliance Assurance Accomplishments Report. http://www.epa.gov/compliance/resources/reports/accomplishments/oeca/fy94accomplishment.pdf

[33] "Is Monsanto Going to Develop or Sell 'Terminator' Seeds?" Monsanto.com. http://www.monsanto.com/newsviews/Pages/terminator-seeds.aspx

[34] Ibid.

[35] Ledford, Heidi. "Seed-patent case in Supreme Court." *Nature,* February 19, 2013. http://www.nature.com/news/seed-patent-case-in-supreme-court-1.12445

[36] http://en.wikipedia.org/wiki/Golden_rice

[37] Gordon, Sarah. "Bill Gates and Monsanto Team Up to Fight World Hunger." *Indiana Public Media*, February 9, 2012. http://indianapublicmedia.org/eartheats/bill-gates-monsanto-team-world-hunger/

[38] Ibid.

[39] Smith, Gar. "Percy Schmeiser vs. Monsanto." *Earth Island Journal*, Autumn 2001.

[40] "Farmers Reporting Farmers – Part 2." Monsanto.com. http://www.monsanto.com/newsviews/Pages/Farmers-Reporting-Farmers-Part-2.aspx

[41] "Seed Police? Part 4." Monsanto.com. http://www.monsanto.com/newsviews/Pages/Seed-Police-Part-4.aspx

[42] *Vernon Hugh Bowman, Petitioner v. Monsanto Company, et al.* United States Court of Appeals for the Federal Circuit Cases Nos. 2010-1068. Docketed September 21, 2011. http://www.supremecourt.gov/Search.aspx?FileName=/docketfiles/11-796.htm

[43] Scahill, Jeremy. "Blackwater's Black Ops." *The Nation,* October 4, 2010. http://www.thenation.com/article/154739/blackwaters-black-ops

[44] "Nation Magazine Story, 'Blackwater Black Ops' Refers to Monsanto and Security Firm" Monsanto.com http://www.monsanto.com/newsviews/Pages/monsanto-blackwater-black-ops.aspx

[45] Vidal, John. "WikiLeaks: US targets EU over GM crops." *The Guardian,* January 3, 2011. http://www.guardian.co.uk/world/2011/jan/03/wikileaks-us-eu-gm-crops

[46] "Ambassador Craig Roberts Stapleton." Embassy of the United States, Paris, France. http://web.archive.org/web/20071130105045/http://france.usembassy.gov/ambassador/default.htm

[47] "Dorothy Walker Stapleton." NNDB. http://www.nndb.com/people/063/000118706/

[48] (Vidal.)

[49] Lean, Geoffrey. "Charles: 'I blame GM crops for farmers' suicides'" *The Independent,* October 5, 2008. http://www.independent.co.uk/environment/green-living/charles-i--blame-gm-crops-for-farmers-suicides-951807.html

[50] Patel, Raj. *Stuffed and Starved.* London: Portobello Books, 2007.

[51] "Report sought on India farm suicides." *BBC News India,* December 21, 2011. http://www.bbc.co.uk/news/world-asia-india-16281063

[52] Stephenson, Wesley. "Indian farmers and suicide: How big is the problem?" *BBC News Magazine*, January 22, 2013. http://www.bbc.co.uk/news/magazine-21077458

[53] Patel, Vikram et. al. "Suicide mortality in India: a nationally representative survey." *The Lancet* 2012; 379: 2343-51

[54] (Stephenson.)

[55] Lerner, George. "Activist: Farmer suicides in India linked to debt, globalization." *CNN*, January 5, 2010. http://www.cnn.com/2010/WORLD/asiapcf/01/05/india.farmer.suicides/index.html

[56] "Vandana Shiva on Farmer Suicides, the US-India Nuclear Deal, Wal-Mart in India and More." *Democracy Now,* December 13, 2006. http://www.democracynow.org/2006/12/13/vandana_shiva_on_farmer_suicides_the

[57] "Indian Farmer Suicide—The Bottom Line." *Beyond the Rows,* Monsanto.com. http://monsantoblog.com/2009/03/26/indian-farmer-suicide-the-bottom-line/

[58] Schwartz, John. "Probe of 3 FDA Officials Sought." *Washington Post,* April 19, 1994. http://legacy.library.ucsf.edu/documentStore/e/n/t/ent92e00/Sent92e00.pdf

[59] "Monsanto, the Government, Monopoly Claims." Monsanto.com. http://www.monsanto.com/food-inc/Pages/monsanto-revolving-door.aspx

[60] Frank, Joshua. "Elena Kagan and Monsanto." *Counterpunch,* May 9, 2010. http://www.counterpunch.org/2010/05/19/elena-kagan-and-monsanto/

[61] Broydo, Laura. "Organic Engineering." *Mother Jones,* March 12, 1998. http://www.pmac.net/lb1.htm

[62] http://en.wikipedia.org/wiki/Tom_Vilsack

[63] Merrigan, Kathleen A., "Principles Driving U.S. Governance of Agbiotech." In Tayor, Iain E. P., ed., *Genetically Engineered Crops: Interim Policies, Uncertain Legislation.* Binghamton: The Haworth Press, 2007.

[64] (Laskawy.)

[65] Cummings, Claire Hope. *Uncertain Peril: Genetic Engineering and the Future of Seeds.* p. 14

[66] http://en.wikipedia.org/wiki/DDT#Use_in_the_1940s_and_1950s

[67] Dracos, Ted. *Biocidal: Confronting the Poisonous Legacy of PCBs.* p. 17

[68] http://en.wikipedia.org/wiki/Monsanto#United_States

[69] Hankinson, Susan et. al., "Circulatory Concentrations of Insulin-like Growth Factor I and Risk of Breast Cancer," *Lancet,* vol. 351, no. 9113 (May 9, 1998), pp. 1393-96.

[70] POSILAC (rBGH) Insert. www.psr.org/assets/pdfs/posilac-insert.doc

[71] Monsanto Letter to Jay Hawkins, Office of Senator Jeffords. December 14, 1998.

[72] Snopes. "Aspartame." http://www.snopes.com/medical/toxins/aspartame.asp

[73] Magnuson BA, Burdock GA, Doull J et al. (2007). "Aspartame: a safety evaluation based on current use levels, regulations, and toxicological and epidemiological studies." Critical Reviews in Toxicology 37 (8): 629–727.

[74] http://en.wikipedia.org/wiki/Agent_Orange

[75] Ibid.

[76] "Veterans' Diseases Associated with Agent Orange." Department of Veterans Affairs Office of Public Health and Environmental Hazards. http://www.publichealth.va.gov/exposures/agentorange/diseases.asp

[77] Fleischer, Doris Zames. *The Disability Rights Movement: From Charity to Confrontation.* Philadelphia: Temple University Press, 2001.

[78] http://en.wikipedia.org/wiki/Agent_Orange#US_veterans_class_action_lawsuit_against_manufacturers

[79] Fawthrop, Tom. "Agent Orange Victims Sue Monsanto." CorpWatch. November 4, 2004. http://www.corpwatch.org/article.php?id=11638

[80] http://en.wikipedia.org/wiki/Dayton_Project

[81] Singh, Salil. "Norman Borlaug: A Billion Lives Saved." AgBioWorld. http://www.agbioworld.org/biotech-info/topics/borlaug/special.html

[82] http://en.wikipedia.org/wiki/Genetically_modified_wheat#Arguments_against_adoption_of_transgenic_wheat

[83] http://en.wikipedia.org/wiki/Genetically_modified_wheat

[84] Union of Concerned Scientists. "Genetic Engineering Has Failed to Significantly Boost U.S. Crop Yields Despite Biotech Industry Claims" http://www.ucsusa.org/news/press_release/ge-fails-to-increase-yields-0219.html

[85] "Can Genetically Modified Crops Feed the World?" *Scientific American,* April 16, 2009. http://www.scientificamerican.com/podcast/episode.cfm?id=can-genetically-modified-crops-feed-09-04-16

[86] World Hunger Education Service. "2012 World Hunger and Poverty Facts and Statistics." http://www.worldhunger.org/articles/Learn/world%20hunger%20facts%202002.htm#Number_of_hungry_people_in_the_world

[87] M. Kummu, H. de Moel, M. Porkka, S. Siebert, O. Varis, P. J. Ward. "Lost food, wasted resources: Global food supply chain losses and their impacts on freshwater, cropland, and fertilizer use." *Science of the Total Environment,* Nov 1, 2012, Vol. 438, pp. 477-489 http://www.sciencedirect.com/science/article/pii/S0048969712011862

[88] (Hart, p. 8-9.)

[89] Smith, Jeffrey. "Spilling the Beans: Unintended GMO Health Risks." Organic Consumers Association. http://www.organicconsumers.org/articles/article_11361.cfm

[90] Eng, Monica. "GMOs should be safety tested before they hit the market says AMA." *Chicago Tribune,* June 19, 2012. http://articles.chicagotribune.com/2012-06-19/features/chi-gmos-should-be-safety-tested-before-they-hit-the-market-says-ama-20120619_1_bioengineered-foods-ama-drug-cosmetic-act

[91] (Palmer.)

[92] R. Mesnage, E. Clair, S. Gress, C. Then, A. Székács, G.-E. Séralini. "Cytotoxicity on human cells of Cry1Ab and Cry1Ac Bt insecticidal toxins alone or with a glyphosate-based herbicide." *Journal of Applied Toxicology,* Feb 15, 2012. http://onlinelibrary.wiley.com/doi/10.1002/jat.2712/abstract

[93] (Palmer.)

[94] NRC. *Environmental effects of transgenic plant: The scope and adequacy of regulation.* Washington DC: National Academy Press. 2002.

[95] Marvier, Michelle and Sabrina West, "Ecological Risk Assessment of GE Crops: Getting the Science Fundamentals Right." In Tayor, Iain E. P., ed., *Genetically Engineered Crops: Interim Policies, Uncertain Legislation.* Binghamton: The Haworth Press, 2007. p. 39

[96] (Brower.)

[97] Randall, Jeff. "Prince Charles warns GM crops risk causing the biggest-ever environmental disaster." *The Telegraph,* August 12, 2008. http://www.telegraph.co.uk/earth/earthnews/3349308/Prince-Charles-warns-GM-crops-risk-causing-the-biggest-ever-environmental-disaster.html

[98] "Hungary Destroys All Monsanto GMO Maize Fields," *International Business Times,* July 22, 2011.

[99] (Winston, p. 232.)

Bibliography

Books

Cook, Guy. *Genetically Modified Language.* New York: Routledge, 2005.

Cummings, Claire Hope. *Uncertain Peril: Genetic Engineering and the Future of Seeds.* Boston: Beacon Press, 2008.

Dracos, Ted. *Biocidal: Confronting the Poisonous Legacy of PCBs.* Boston: Beacon Press, 2010.

Ellstrand, Norman C. *Dangerous Liasons? When Cultivated Plants Mate With Their Wild Relatives.* Baltimore: Johns Hopkins University Press, 2003.

Fleischer, Doris Zames. *The Disability Rights Movement: From Charity to Confrontation.* Philadelphia: Temple University Press, 2001.

Hankinson, Susan et. al., "Circulatory Concentrations of Insulin-like Growth Factor I and Risk of Breast Cancer," *Lancet,* vol. 351, no. 9113 (May 9, 1998).

Hart, Kathleen. *Eating in the Dark: America's Experiment With Genetically Engineered Food.* New York: Pantheon Books, 2002.

Lurquin, Paul F. *The Green Phoenix: A History of Genetically Modified Plants.* New York: Columbia University Press, 2001.

Nestle, Marion. *Safe Food: The Politics of Food Safety.* Berkeley: University of California Press, 2010. p. 279

Patel, Raj. *Stuffed and Starved.* London: Portobello Books, 2007.

Tayor, Iain E. P., ed., *Genetically Engineered Crops: Interim Policies, Uncertain Legislation.* Binghamton: The Haworth Press, 2007.

Thomson, Jennifer A. *Seeds for the Future: The Impact of Genetically Modified Crops on the Environment.* Cornell University Press, 2007.

Winston, Mark L. *Travels in the Genetically Modified Zone.* Cambridge: Harvard University Press, 2002

Articles and Reports

Brower, Lincoln. "The Monarch and the Bt Corn Controversy." *Orion*, Spring 2001. http://www.orionmagazine.org/index.php/articles/article/85

Broydo, Laura. "Organic Engineering." *Mother Jones,* March 12, 1998. http://www.pmac.net/lb1.htm

"Can Genetically Modified Crops Feed the World?" *Scientific American,* April 16, 2009. http://www.scientificamerican.com/podcast/episode.cfm?id=can-genetically-modified-crops-feed-09-04-16

Chan, Amanda. "'Bullet-Proof Skin', Made With Spider Silk and Goat's Milk, Created by Scientists." *Huffington Post,* August 18,

2011. http://www.huffingtonpost.com/2011/08/18/bullet-proof-skin-spider-silk_n_930389.html

"Continuing the Green Revolution: The corporate assault on the security of the global food supply." Massachusetts Institute of Technology: *The Thistle.* Vol. 13, No. 4. June/July 2001. http://web.mit.edu/thistle/www/v13/4/food.html

Eng, Monica. "GMOs should be safety tested before they hit the market says AMA." *Chicago Tribune,* June 19, 2012. http://articles.chicagotribune.com/2012-06-19/features/chi-gmos-should-be-safety-tested-before-they-hit-the-market-says-ama-20120619_1_bioengineered-foods-ama-drug-cosmetic-act

Fawthrop, Tom. "Agent Orange Victims Sue Monsanto." CorpWatch. November 4, 2004. http://www.corpwatch.org/article.php?id=11638

Frank, Joshua. "Elena Kagan and Monsanto." *Counterpunch,* May 9, 2010. http://www.counterpunch.org/2010/05/19/elena-kagan-and-monsanto/

Gillam, Carey. "Monsanto, DuPont strike $1.75 licensing deal, end lawsuits." Reuters. March 26, 2013. http://www.reuters.com/article/2013/03/26/us-monsanto-dupont-gmo-idUSBRE92P0IK20130326

Gordon, Sarah. "Bill Gates and Monsanto Team Up to Fight World Hunger." *Indiana Public Media*, February 9, 2012. http://indianapublicmedia.org/eartheats/bill-gates-monsanto-team-world-hunger/

"Hungary Destroys All Monsanto GMO Maize Fields," *International Business Times,* July 22, 2011.

Knowles, David. "Opponents of genetically modified organisms in food, or GMOs, rail against provision that would limit the courts' ability to stop food producer Monsanto from growing crops later deemed potentially hazardous." *New York Daily News,* March 25, 2013. http://www.nydailynews.com/news/national/food-oversight-curbs-spending-bill-outrage-article-1.1298967

Ledford, Heidi. "Seed-patent case in Supreme Court." *Nature,* February 19, 2013. http://www.nature.com/news/seed-patent-case-in-supreme-court-1.12445

Lerner, George. "Activist: Farmer suicides in India linked to debt, globalization." *CNN,* January 5, 2010. http://www.cnn.com/2010/WORLD/asiapcf/01/05/india.farmer.suicides/index.html

Palmer, Roxanne. "GMO Health Risks: What the Scientific Evidence Says." *International Business Times,* March 30, 2013. http://www.ibtimes.com/gmo-health-risks-what-scientific-evidence-says-1161099

Pinder, Terry. "Monsanto is a bad corporation but that GM Maize study is dodgy as heck." *Daily Kos,* September 20, 2012. http://www.dailykos.com/story/2012/09/20/1134246/-Monsanto-is-a-bad-corporation-but-that-GM-Maize-study-is-dodgy-as-heck

Randall, Jeff. "Prince Charles warns GM crops risk causing the biggest-ever environmental disaster." *The Telegraph,*

August 12, 2008. http://www.telegraph.co.uk/earth/earthnews/3349308/Prince-Charles-warns-GM-crops-risk-causing-the-biggest-ever-environmental-disaster.html

"Report sought on India farm suicides." *BBC News India,* December 21, 2011. http://www.bbc.co.uk/news/world-asia-india-16281063

Scahill, Jeremy. "Blackwater's Black Ops." *The Nation,* October 4, 2010. http://www.thenation.com/article/154739/blackwaters-black-ops

Schneider, K. "Faking It: The Case Against Industrial Bio-Test Laboratories." *The Amicus Journal,* Spring 1983. http://planetwaves.net/contents/faking_it.html

Schwartz, John. "Probe of 3 FDA Officials Sought." *Washington Post,* April 19, 1994. http://legacy.library.ucsf.edu/documentStore/e/n/t/ent92e00/Sent92e00.pdf

Singh, Salil. "Norman Borlaug: A Billion Lives Saved." AgBioWorld. http://www.agbioworld.org/biotech-info/topics/borlaug/special.html

Snopes. "Aspartame." http://www.snopes.com/medical/toxins/aspartame.asp

Stephenson, Wesley. "Indian farmers and suicide: How big is the problem?" *BBC News Magazine,* January 22, 2013. http://www.bbc.co.uk/news/magazine-21077458

Roberts, Jeanne. "The Real Victor in Iraq: Monsanto," in *The*

Panelist. http://www.thepanelist.net/opinions-culture-10084/1252-the-real-victor-in-iraq-monsanto

Smith, Gar. "Percy Schmeiser vs. Monsanto". *Earth Island Journal*, Autumn 2001.

Smith, Jeffrey. "Spilling the Beans: Unintended GMO Health Risks" Organic Consumers Association. http://www.organicconsumers.org/articles/article_11361.cfm

"The goats with spider genes and silk in their milk," BBC News. http://www.bbc.co.uk/news/science-environment-16554357

"Vandana Shiva on Farmer Suicides, the US-India Nuclear Deal, Wal-Mart in India and More." *Democracy Now,* December 13, 2006. http://www.democracynow.org/2006/12/13/vandana_shiva_on_farmer_suicides_the

Vidal, John. "WikiLeaks: US targets EU over GM crops." *The Guardian,* January 3, 2011. http://www.guardian.co.uk/world/2011/jan/03/wikileaks-us-eu-gm-crops

Statements, Press Releases, Letters and Documents

Ambassador Craig Roberts Stapleton." Embassy of the United States, Paris, France. http://web.archive.org/web/20071130105045/http://france.usembassy.gov/ambassador/default.htm

"Dorothy Walker Stapleton." NNDB. http://www.nndb.com/people/063/000118706/

EPA FY1994 Enforcement and Compliance Assurance Accomplishments Report. http://www.epa.gov/compliance/resources/reports/accomplishments/oeca/fy94accomplishment.pdf

"Farmers Reporting Farmers – Part 2." Monsanto.com. http://www.monsanto.com/newsviews/Pages/Farmers-Reporting-Farmers-Part-2.aspx

Federal Register Volume 66, Number 98. May 21, 2001. pp. 28045-28046 http://www.gpo.gov/fdsys/pkg/FR-2001-05-21/html/01-12966.htm

"Indian Farmer Suicide—The Bottom Line." *Beyond the Rows,* Monsanto.com. http://monsantoblog.com/2009/03/26/indian-farmer-suicide-the-bottom-line/

"Is Monsanto Going to Develop or Sell 'Terminator' Seeds?" Monsanto.com. http://www.monsanto.com/newsviews/Pages/terminator-seeds.aspx

Monsanto Letter to Jay Hawkins, Office of Senator Jeffords. December 14, 1998.

"Monsanto, the Government, Monopoly Claims." Monsanto.com. http://www.monsanto.com/food-inc/Pages/monsanto-revolving-door.aspx

"Monsanto Statement Regarding Farmer Assurance Provision in H.R.933." Monsanto.com. http://www.monsanto.com/newsviews/Pages/statement-regarding-farmers-assurance-provisions.aspx

"Nation Magazine Story, 'Blackwater Black Ops' Refers to Monsanto and Security Firm" Monsanto.com http://www.monsanto.com/newsviews/Pages/monsanto-blackwater-black-ops.aspx

Rogers, Harold, Hon. Letter to Honorable Norm Dicks. June 12, 2012. Monsanto.com. http://www.monsanto.com/newsviews/Documents/Letter%20to%20House%20for%20Section%20733.pdf

Roundup Pro® Herbicide Complete Directions for Use. EPA Reg. No. 524-475. http://www.afpmb.org/sites/default/files/pubs/standardlists/labels/6840-01-108-9578_label_roundup_pro.pdf

"Seed Police? Part 4." Monsanto.com. http://www.monsanto.com/newsviews/Pages/Seed-Police-Part-4.aspx

"Toxic pollen from widely planted, genetically modified corn can kill monarch butterflies, Cornell study shows." Ithaca: Cornell University. May 19, 1999. http://web.archive.org/web/20130323114050/http://www.news.cornell.edu/releases/May99/Butterflies.bpf.html

Union of Concerned Scientists. "Genetic Engineering Has Failed to Significantly Boost U.S. Crop Yields Despite Biotech Industry Claims" http://www.ucsusa.org/news/press_release/ge-fails-to-increase-yields-0219.html

Vernon Hugh Bowman, Petitioner v. Monsanto Company, et al. United States Court of Appeals for the Federal Circuit Cases Nos. 2010-1068. Docketed September 21, 2011. http://www.supremecourt.gov/Search.aspx?FileName=/docketfiles/11-796.htm

Reports and Studies

Hankinson, Susan et. al., "Circulatory Concentrations of Insulin-like Growth Factor I and Risk of Breast Cancer," *Lancet,* vol. 351, no. 9113 (May 9, 1998), pp. 1393-96.

Magnuson BA, Burdock GA, Doull J et al. (2007). "Aspartame: a safety evaluation based on current use levels, regulations, and toxicological and epidemiological studies". Critical Reviews in Toxicology 37 (8): 629–727.

M. Kummu, H. de Moel, M. Porkka, S. Siebert, O. Varis, P. J. Ward. "Lost food, wasted resources: Global food supply chain losses and their impacts on freshwater, cropland, and fertilizer use." *Science of the Total Environment,* Nov 1, 2012, Vol. 438, pp. 477-489 http://www.sciencedirect.com/science/article/pii/S0048969712011862

NRC. *Environmental effects of transgenic plant: The scope and adequacy of regulation.* Washington DC: National Academy Press. 2002.

Patel, Vikram et. al. "Suicide mortality in India: a nationally representative survey." *The Lancet* 2012; 379: 2343-51

R. Mesnage, E. Clair, S. Gress, C. Then, A. Székács, G.-E. Séralini. "Cytotoxicity on human cells of Cry1Ab and Cry1Ac Bt insecticidal toxins alone or with a glyphosate-based herbicide." *Journal of Applied Toxicology,* Feb 15, 2012. http://onlinelibrary.wiley.com/doi/10.1002/jat.2712/abstract
Séralini GE, Clair E, Mesnage R, Gress S, Defarge N, Malatesta M, Hennequin D, de Vendômois JS. "Long term toxicity of a Roundup herbicide and a Roundup-tolerant

genetically modified maize." http://www.ncbi.nlm.nih.gov/pubmed/22999595

—. "Answers to critics: Why there is a long term toxicity due to a Roundup-tolerant genetically modified maize and to a Roundup herbicide." http://www.ncbi.nlm.nih.gov/pubmed/23146697

The Pew Trust. *Three Years Later: Genetically Engineered Corn and the Monarch Butterfly Controversy.* http://www.pewtrusts.org/uploadedFiles/wwwpewtrustsorg/Reports/Food_and_Biotechnology/vf_biotech_monarch.pdf

World Hunger Education Service. "2012 World Hunger and Poverty Facts and Statistics." http://www.worldhunger.org/articles/Learn/world%20hunger%20facts%202002.htm#Number_of_hungry_people_in_the_world

Jason Louv is an independent journalist who is not funded by anybody, thank you kindly. He has written for *Esquire Online, Motherboard (VICE), Humanity Plus* and many others. You can follow him on Twitter at @jasonlouv, on Facebook at www.facebook.com/ultraculturegate and on the Web at www.ultraculture.org.

ultraculture.org

24002907R00037

Made in the USA
Lexington, KY
02 July 2013